WEST BROMWICH BEFORE THE INDUSTRIAL REVOLUTION

by

D. DILWORTH, M.A., B.Sc.

A BLACK COUNTRY SOCIETY
PUBLICATION

First impression 1973

ISBN 0 9501197 9 2

Printed by Reliance Printing Works, Birmingham Street, Halesowen, Worcs.

CONTENTS

Abbreviations

P.R.O. .... Public Record Office

S.Hist.Coll. (S.H.C.) .... Staffordshire Historical Collection

W.B.M.C.R. .... West Bromwich Manorial Court Rolls

W.B.R.L. .... West Bromwich Reference Library

W.S.L. .... William Salt Library, Stafford

Foreword

THE excavated foundations of some stronghold, the broken shard, the decayed 'lares and penates' of some once-proud owner, the strip lynchet; such are the clues for which the archaeologist searches, and such are the finds on which he builds his reconstruction of a bygone civilisation. In open country, surface markings, often revealed for the first time by aerial photograph, suggest the probable fruitful site for investigation. Here, effort and care are the main pre-requisites. In an urban area the problem is far more complicated and even where modern city is known to stand on Roman settlement, it is usually necessary to wait until commercial interest decides on re-development before work can be carried out, and then usually only in a piecemeal fashion.

But not every 20th century Leicester covers a Roman Ratae, and if it is difficult to determine the extent and conditions of an earlier urban community when submerged beneath present-day bricks and concrete, how much greater is the problem of discovering the details of earlier rural settlement in a similarly built-up area? The outlines of field and wood, footpaths, the sites of wattle and daub structures, leave little enough trace in any case, and are usually completely obliterated in areas subjected to 18th and 19th century industrial growth.

Such is the case in the Black Country towns. Nevertheless, the historical geographer need not despair of making a reasonable reconstruction of the pre-industrial era even here. A detailed search may reveal the evidence of an old farmhouse, barn or mill. A modern road often follows an earlier path or hedge line. An early tithe map or property plan can be of great value, though few of these exist for this part of the world prior to the 19th century.

Place names are often indicative and helpful. For the rest, a mass of documentary evidence in the shape of Manorial Records and Rent Rolls, Terriers, Deeds of Sale or Lease, Assize Records and Parish Registers exists, and while incomplete in themselves, they provide a host of clues and details for research. It is in this way and from such sources that I have attempted to trace some seven centuries in the life of an agricultural West Bromwich. Not the West Bromwich of the present-day boundaries, but that of the original manorial holding as defined by Reeves in his History of West Bromwich, 1836.

April, 1973. D. DILWORTH.

Topography

THE one-time Manorial lands of West Bromwich were bordered by the Upper Tame Valley to the south-west, west, north-west and north-east. To the south-east the boundary followed no marked feature, but crossed the neck of a spur. The area within these bounds formed a part of the Birmingham Plateau, a broad spur stretching out in a north-west and northerly direction, the broad flat top falling from just over 500 ft. in the south-east to a little below that figure at the northern end. The smooth outline of this plateau has been broken by the small tributaries of the Tame. The 'Hobnail Brook', in the northern part of the area, flows northward forming a broad shallow valley. Two nameless streams in the eastward part flow eastward to the river, draining a very broad corrie-shaped indentation in the higher land. For the most part the slopes bounding our plateau are very gentle. For most of the western side they are of the order of 1 in 75. Though they steepen to something like 1 in 35 in the north; only at one point (990939) does the higher land approach the river, giving a gradient of 1 in 14. This, being opposite to the 'island' of Wednesbury, forming the narrowest part of the river valley, it is not surprising that the route of the turnpike road was so chosen that it would pass through this point. Though the gradient was steeper, the broad marshy stretches of the valley would be avoided. As might be expected the slopes overlooking the deeper valley to the east are somewhat steeper. They are generally of a concave nature with gradients of about 1 in 10 to 1 in 15 at the upper part of the slopes.

For a small stream, the upper part of a river system, the Tame has little fall. From the 450 ft. contour to the 400 ft. is a distance of $2\frac{1}{2}$ miles, a gradient of 1 in 264, while there is an even gentler fall over the next 50 ft., the distance required being $4\frac{3}{4}$ miles, a gradient of 1 in 502. There is nothing visible on the surface today to suggest a marked change of gradient at any point, for in many places the course of the stream has been altered and regulated to adjust it to the man-made requirements of canals, dams and sewage plants. There is, however, a possibility that there was something of a nick point at the narrowest part of the valley above-mentioned. The extreme width and lack of relief of the valley

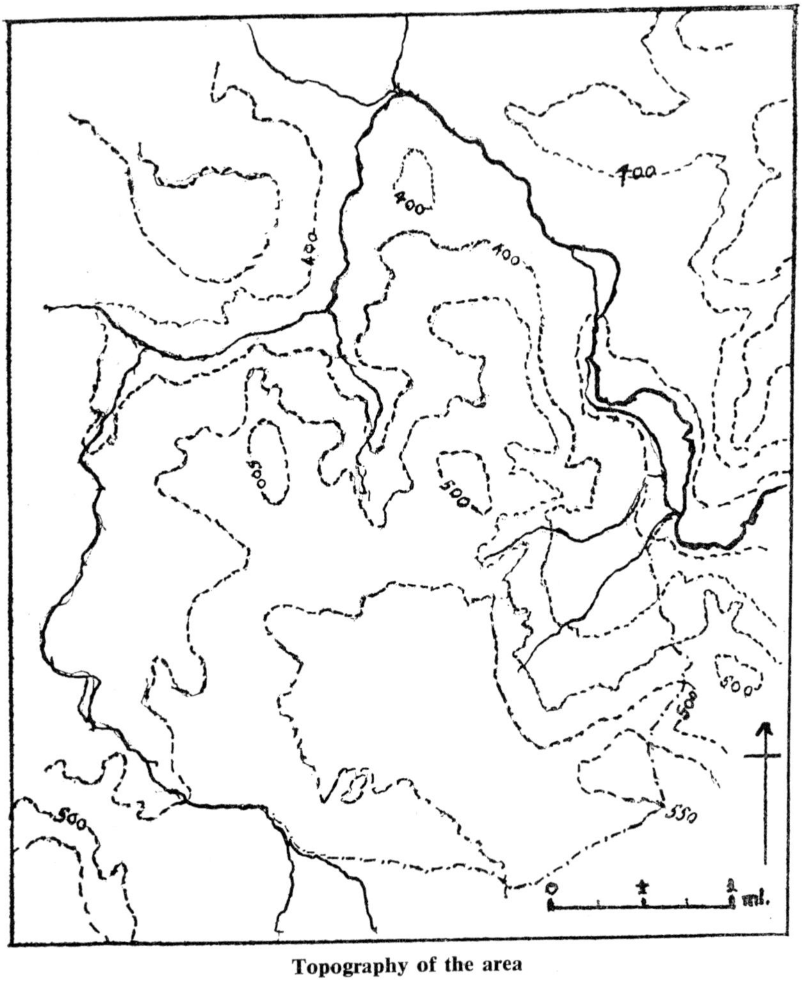

Topography of the area

bottom suggests that in this part of its course the river is draining two very shallow basins. The upper one lies upstream of the 'Wednesbury narrows' and is very broad and flat. At one part it is at least two miles across the valley between the two occurrences of the 450 ft. contour without any variation exceeding 25 ft. From the 'narrows' the stream flows northward to the point (005963) where it receives its largest tributaries, the Willenhall Brook and the Walsall Water or Tame. At this point there opens out to the northward a quite extensive shallow basin some six miles by three. The Tame, however, here turns in a south-easterly direction, following an extension of this basin which continues in this direction for about another three miles with a width of about a mile. This basin is closed by another narrowing at Hamstead (048928) which is just beyond our area.

Although the Tame only receives three small tributaries from the West Bromwich area within this bend, it is rather better served by the acquisition of a number of left bank streams. Of these, the Tipton Brook, the Leabrook, the Willenhall Brook, the Walsall Tame and the Fullbrook are the most important. It is interesting to note that there appear to have been few springs occurring on the western side of our plateau, while there were several on the eastern side. It is difficult at this date to be certain on this point, when much of the area is now built over, and when adequate early maps are lacking.

Along the eastern part of the valley there occur several small terraces at about the 420 ft. level. They appear for the most part to be covered with gravel, but whether they are true terrace gravels or are materials washed down from the drift deposits above, is not certain.

Climate

THE climate of West Bromwich is essentially that of the West Midlands, with a rainfall lying between that of the drier eastern counties and that of the wetter west, with an annual temperature range somewhat high for England as a whole; a climate dominated by the passage of successive cyclonic features.

The climate of the area has been admirably described in considerable detail by Miss Saward (i). Her findings may be briefly summarised as follows:—

Winds. The only records available were those of Edgbaston Observatory. As might have been predicted, winds from a south-westerly direction predominated, with south, north-west and westerly sectors, in that order following. The local effects of building on both direction and velocities had not been assessed.

Rainfall. The rainfall, which was mainly frontal in origin, varied from 22 to 35 inches over the district as a whole, reflecting generally the relief of the land. Our area, the old parish of West Bromwich, lying to the north-west of the axial ridge of the Birmingham Plateau, would be in something of a rain-shadow area. Miss Saward's rainfall map, part of which is reproduced here, shows the precipitation as varying from about 29 inches in the south-west to about 25 inches in the east in the Tame valley. The outstanding fact regarding rainfall in this district is the considerable variation that can occur in monthly, seasonal or annual amounts. This is well illustrated in her diagram based on Edgbaston figures.

These figures show the annual distribution with the heaviest rainfall occurring in the late autumn, with a maximum in October and a secondary maximum in August, February and March being the driest months. Summer thunderstorms are not infrequent.

Temperature. Temperature ranges and characteristics varied in step with other areas of England and Wales, depending on the prevalence of cyclonic conditions. Edgbaston figures showed a considerable variation in the monthly averages from one year to another. The annual range of monthly mean temperatures was of the order of 23°F. Appreciable differences of temperature occur in different localities and reflect proximity to largely built-up areas.

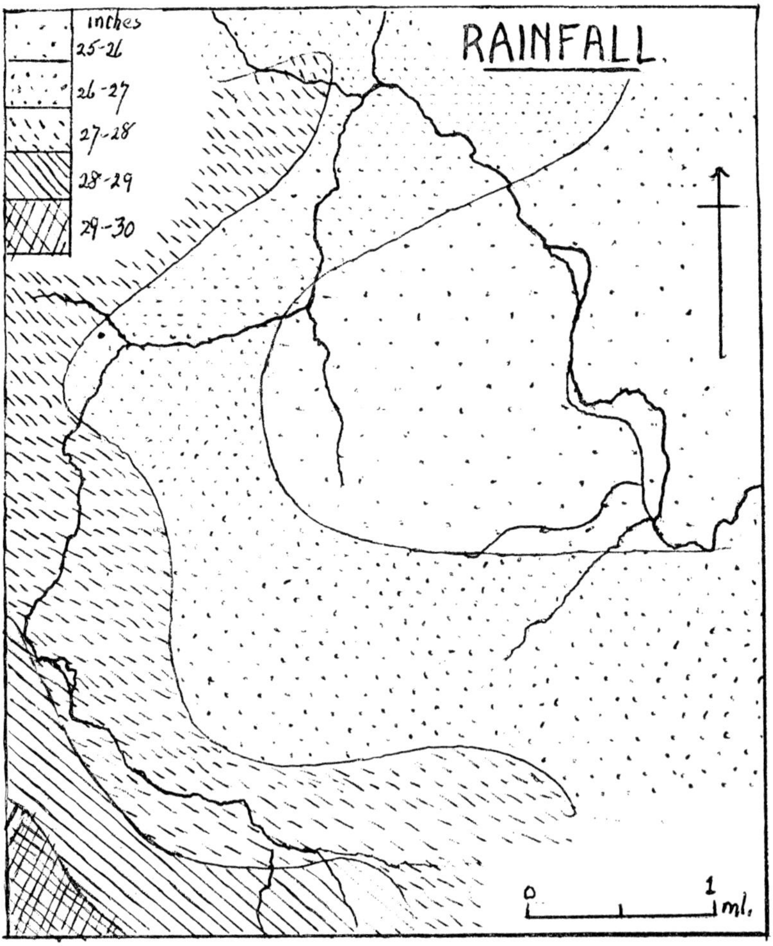

After B. Saward

Average Annual Rainfall

Fog, Haze and Sunshine. Atmospheric pollution was an important factor, industrial haze being common particularly under conditions of temperature inversion. The location of fog was greatly affected by wind direction, the commonest types of fog being radiation and smoke fogs. Sunshine figures, generally low compared with other inland sites, varied considerably with the locality within the area. The atmospheric pollution of the Black Country considerably reduced the visibility and hence the effective sunshine figures.

Any variation of the climate of West Bromwich from the general pattern for the Birmingham area as outlined above is likely to be of interest mainly to a micro-climatologist. Records over a considerable period from a number of well distributed stations would be needed for a detailed comparison and such do not exist. Such data as are available (ii) though insufficient to establish any real difference are worthy of consideration in that they may at least reveal certain trends.

West Bromwich lies to the north-east and east of Smethwick and Oldbury respectively, i.e., the highest part of the Tame valley, an area of particularly heavy industrialisation. Inevitably the prevailing south-west and west winds bring with them a more than usually large amount of atmospheric pollution. This must not only reduce the amount of effective sunshine received, an effect which Miss Saward indicated, but it must also act as a blanket reducing radiation effects when the wind is in those directions. This in turn would be expected to affect maximum and minimum temperatures.

Rainfall can more easily be assessed. The average rainfall over a period of fifteen years was 27.21 ins. which is slightly more (possibly 1 in.) than would be expected from the map. The distribution follows the same pattern with a late autumn and winter maximum, October being the autumn month with the highest average. Again February is the month with the lowest monthly average. The secondary maximum which appears to be in August for the whole area, here occurs in July, yielding in fact the highest monthly average of the year, 2.80 ins. This is undoubtedly due to the incidence of particularly heavy summer thunderstorms during the period recorded. There was a fall of 1.02 ins. on July 21st, 1930, 1.38 ins. on the 2nd and 1.95 ins. on the 9th of the month in 1936 and 2.40 ins. on 15th July, 1937. The same variability of monthly, seasonal and yearly amounts recorded at Edgbaston is in evidence here. The fall in February, 1965, was 0.50 ins. while in February of the following year it was 3.81 ins. Similarly the fall for October, 1965, was only 0.67 ins. compared with 6.89 ins. for October of 1967.

Yearly totals also vary widely from 22.90 ins. in 1934 to

32.25 ins. in 1966. We have no records of rainfall in this area for past centuries (Edgbaston figures are only available since 1891) and so must assume that there has been no marked change in amount or distribution pattern. However, the wholesale deforestation of the area, largely in the 15th to 18th centuries, would greatly reduce the amount of soil-retained precipitation, through increase both of run-off and of evaporation.

The pattern of temperature readings for West Bromwich follows closely those of Edgbaston, both in the matter of variability and range. The most recent figures available (iii), unfortunately over a period of five years (1963 to 1967) only, show February figures varying from 30.0°F. in 1963 to 41.5°F. in 1966, and August figures which vary from 55.5°F. in 1963 to 61.5°F. in 1964. The annual range of mean monthly temperatures was slightly over 23.0°F. These figures are probably severely biased by the fact that this short period included the very unusually cold spell of the 'big freeze' of January, February and early March of 1963. Given more truly representative figures, we should probably find that the annual range is something of the order of 21.0°F.

Has there been any material change in temperature, severity of winters or frequency of late frosts during the past two centuries? These are questions to which it is not possible to obtain definite answers, but nevertheless it is interesting to note that when Sir Charles Clarke bought the Manor of West Bromwich in 1720, he established extensive orchards on adjacent ground which lies at a height of about 470 ft. Included in his original 'shopping list' were grape vines, peaches, nectarines and apricots. There is reason to believe that this venture met with some success for similar purchases were made over the next forty years (iv). In 1760 he added a fig tree to his purchases! Was he too optimistic?

The statistics for this period as set out by Professor Manley do not suggest that there have been any marked changes (v).

(i) B. Saward in *Birmingham and its Regional Setting* (1950), p. 47.
(ii) Recordings taken at Dartmouth Park, West Bromwich.
(iii) Recordings taken at George Salter Boys' School, West Bromwich.
(iv) M. Willett, *History of West Bromwich* (1882).
(v) G. Manley in *Quarterly Journal of the Royal Meteorological Society*, Vol. 79, No. 340, p. 256.

Geology

THIS is not the place to give a detailed geological history of West Bromwich. Very full accounts of the geology of the Midlands will be found in the Geological Survey Memoirs, the Central England Handbook of the Geological Survey, or in Professor L. J. Wills' Palaeogeography of the Midlands. It will suffice for our purpose to examine a generalised distribution of the strata of the area.

Almost the whole of the solid geology consists of carboniferous deposits lying unconformably on the Silurian platform. The only pre-carboniferous deposits exposed within our boundary occur in the Bustleholme Mill area in the north-east, where the southernmost tip of the Walsall Silurian inlier is to be found.

The lower carboniferous series are absent here, and the rest of our solid geology ranges from the productive middle coal measures to the barren upper carboniferous through the Etruria Marl, Halesowen beds, Keele beds and the Enville beds. The eastern boundary fault of the South Staffordshire coalfield crosses our area from the Silurian inlier in a south-westerly direction, with the downthrow on the eastern side. This provides us with what Professor Wills (i) calls the exposed and the half-hidden coalfields. To the west of the fault we have the South Staffordshire coalfield horst, with the corresponding graben to the east.

Very little of the productive middle coal measures are actually exposed within our area, a fact which delayed the development of mining and industry in the area and preserved West Bromwich as a predominantly agricultural area until the 19th century. The first pit of which we know the exact position, the Ball's Hill Colliery, was sunk in 1801. Prior to this, in 1719, a 'very good colemine' was being worked in the same area by John Butler of Wednesbury (ii) but there occur no other references during that century. The only areas of exposure occur in the Gold's Hill, Leabrook area of the north-west, two small areas at Friar Park in the north, and a very small area at Wigmore in the north-east.

There is little doubt that these shallow coals were worked from an early date, but when is not known. In July, 1657, the name of John Killiam, 'cole worker', appears in the West Brom-

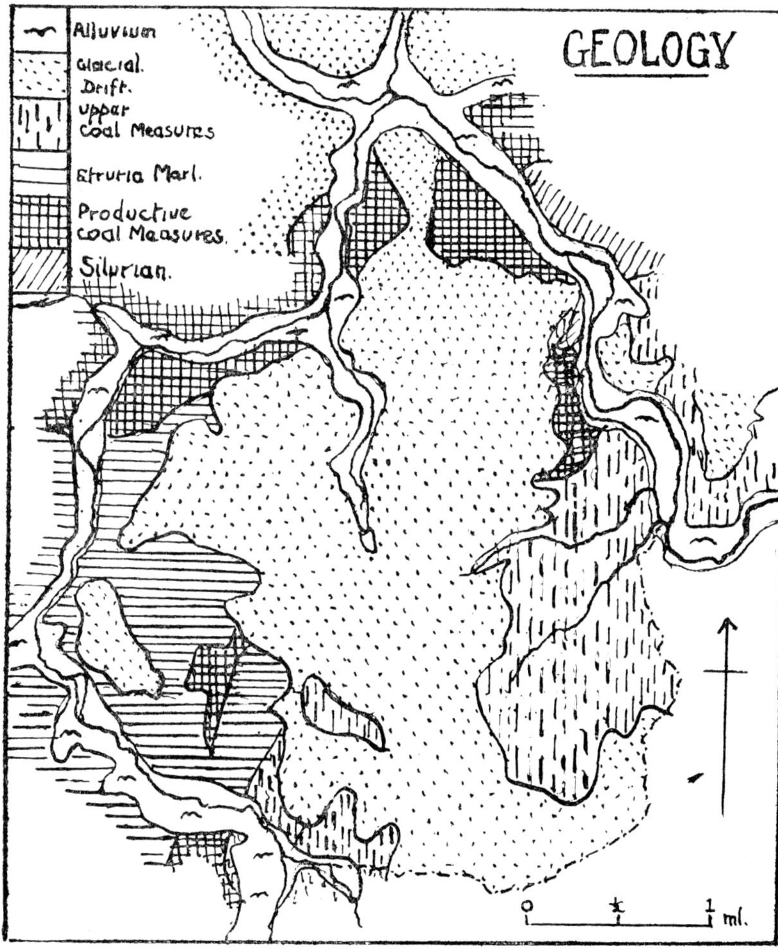
Alluvium
Glacial. Drift.
upper Coal Measures
Etruria Marl.
Productive coal Measures.
Silurian.
GEOLOGY
0
½
1 ml.

wich parish register. In 1660, Brome Whorwood granted a 21 year lease to Thomas Nurthall (iii) of Bald's Hill, the Hayes and Hayes Meadow, reserving the rights to mine stone and ' cole '. It is probable that this coal was in demand to replace the disappearing wood-charcoal in the growing cottage iron industries. It had been enacted as early as 1606 (iv) that ' no inhabitant in this manor, using the trade, mystery or occupation of a smith, naylor or bucklemaker shall set or let any stall or stock to any person or persons of the same trade or occupation to work upon, contrary to the statute, under a penalty of 6s. 8d.'

Similarly, there were penalties for giving shelter to strangers from outside the parish. Despite these embargoes the number of ' immigrant ' settlers engaged in these trades steadily increased throughout the 17th century. At the Manorial Court of 1687 (v) there were no fewer than 71 defendants accused of encroaching on the wastes by the building of a cottage and in 26 cases an adjoining workshop. According to Reeves, the north-western area of exposed coal was already worked out by 1750. There is evidence that some coal was got near to Wigmore, for there is still a small spoil heap now grass covered.

The thickness of overburden, together with a general dip to the south-east, necessitated deeper shafts as we move away from the small exposed areas. Even in the west at 987907, when Izons sank a pit c. 1820 in their Barn field, they had to go down 200 ft. through the Etruria marl to the 10 yd. seam (vi). Beyond the boundary fault only two mines have been opened up, Hamstead, about 1,878 ft. deep, and the Jubilee shaft of the Sandwell Park Co., about 1,480 ft. deep.

The outcrops of the Etruria marl proved of great local importance in providing the raw material for the considerable number of brick-works which sprang up in the 19th century. The upper coal measures, the Keele and Enville beds are mainly deep red sandstones, often very soft, with thin layers of much harder grey-green sandstones. They were, in January, 1968, very well exposed in the making of the new cutting for the M5 extension at Newton Road (022930).

Much of the area over 400 ft. is covered with glacial drift which for the most part consists of light sand and quartzite pebbles, though there are some patches of a deep red boulder clay.

Reeves (vii) states that the soil was generally a rich sandy loam except on the summits of the hills. Sufficient agricultural land still remains in the eastern part of the area for us to observe the accuracy of his statement on the loamy nature of the soils on the slopes. The poorer nature of the hilltop soils was due to the sandy drift material which, if we may compare with similar areas in the Midlands, was quite probably heavily podsolised. This, together with the patches of boulder clay, would account for the number of small marshy areas and natural pools which we know to have existed. Along the bottom of the Tame valley was a wide strip of alluvial soil which was, however, until a few years ago, subject to periodic flooding.

(i) L. J. Wills in *Birmingham and its Regional Setting* (1950), p. 18.
(ii) West Bromwich Rent Roll, 1719.
(iii) M. Willett, *History of West Bromwich* (1882), p. 220.
(iv) E. Lissimore, ed. *Manorial Court Rolls of West Bromwich,* 17th Oct. 4 Jas. I. p. 27.
(v) E. Lissimore, op. cit., 15 Oct. 3 Jas. II. p. 171.
(vi) W. E. Jephcott, A West Bromwich scrapbook in West Bromwich Reference Library.
(vii) J. Reeves, *History of West Bromwich* (1836), p. 95.

Vegetation

ANY attempt to depict the original natural vegetation of a present-day industrial area must rely on a variety of sources, on the descriptive powers of former writers, on references in official documents, and possibly to some extent on local place names. Drawing on all of these, the present writer has attempted to put the clock back.

Local names of a vegetational connotation, which have not been recently thought up by a local council, are surprisingly numerous. Oak trees, possibly because of their durability, achieved importance as boundary marks or reference points. Such were the Meare Oak (i) (030913), the Three Mile Oak (014904) and Harvil's Oak (989932), although the changing of the name of the latter to Harvil's Hawthorne, during the 18th century, hints at the fate of the forest trees in that period, and of their replacement by a ' second-grade ' vegetation.

From 1592, the Turton family lived at ' The Oak ', still known as the ' Oak House '. We know that in former times this house was surrounded by oak trees which were felled about 1767 to supply lock gate timber for the canal at that time being constructed by Brindley (ii). We know that heathland was abundant and the names of Hateley Heath (000936) and Heath Lane indicate one such locality. The Lyng (003908), a name given to an area south of the old turnpike road, tells its own story. The very name Bromwich (or Bromwic) suggests a Saxon hamlet in a countryside of broom, gorse and heathland. The name Hollyhedge Road (012922), given by the Enclosure Act of 1802 (iii) to a road which followed in part the boundary between two of the old Open Fields, denotes the nature of that former boundary.

Written references to the vegetation are somewhat scrappy. The Domesday record, which shows Bromwic as being in Northamptonshire with others of the FitzAnsculph properties, shows a large amount of woodland, seven leagues by half a league. If we take this measurement to be 12 furlongs (iv) this could mean an area of up to 5,000 acres. It is impossible, of course, to know how these measurements were taken, but undoubtedly a high percentage of this manor was wooded. Comparatively little, some

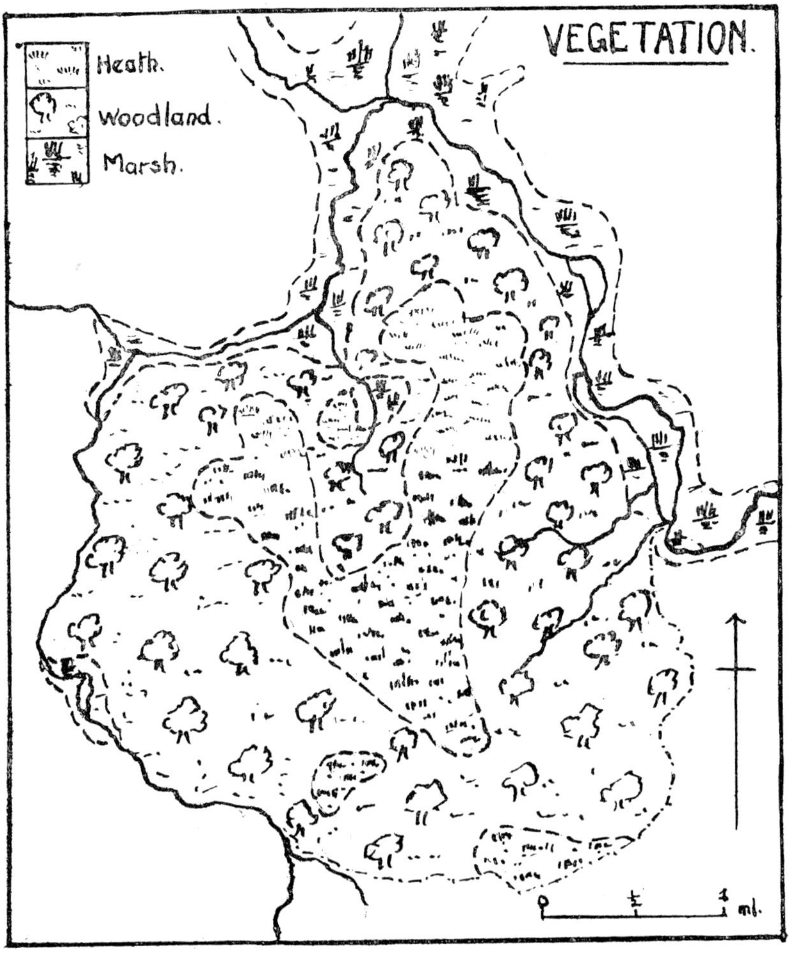

VEGETATION.
Heath.
Woodland.
Marsh.
0
½
1 ml.

400-500 acres probably, was under cultivation. Unfortunately, the record gives no account of waste or heathland here. From this time, the end of the 11th century onwards, there was a steady clearance of woodland. A charter of Gervase Pagnel in 1175, confirming a gift of land to the Priory of Sandwell (v), mentions among other rights that of ' lignum ad focum ' (firewood). Other areas were cleared in order to establish agricultural ground.

During the 13th and 14th centuries we find numerous references to the sale or lease of fields, particularly in the Fynch-path (Hill Top) to Wednesbury Bridge area (vi). By mid-16th century the clearance had certainly extended to the Black Lake, Great Bridge and Greets Green areas, in which much of the cleared land was alienated from manorial possession. Nearer to those centres of the West Bromwich community, the Manor House and the church, the open common fields were early established. A Deed of 22 Ed. III (1348) (vii) refers to Wigmorefield. However, apparently trees could not be felled indiscriminately. In 1573 (viii) a jury was instructed to investigate the felling of seven oak trees in Saltweye Lane. There was a similar case in 1627. Despite this progressive deforestation there was sufficient remaining in 1796 for Pitt (ix) to say that Sandwell Park and other estates were well stocked with all kinds of timber, particularly oak.

Of the heathlands, the largest was the one known as West Bromwich Heath. The original bounds of this cannot be traced, but despite considerable whittling of its area, largely in the 17th and 18th centuries, it was still over a mile in length and covered some 387 acres at the time of the enclosure (x). From the 14th century there are many references to it in local documents such as a Deed of 1637, which mentions a lane from West Bromwich Heath to Wednesbury (xi).

In 1609 (xii), George Symcoxe acknowledged his liability to the manor for a cottage by the heath. During the latter half of the 17th century there were serious invasions of the heath. In 1685 (xiii) no fewer than 71 people were fined for encroachment on the lord's waste, i.e., by the building of cottages or in some cases workshops, since the last count. Many of these, though probably not all, were at the various ' Ends ' of the heath. Hateley Heath, much smaller, being only about ten acres in extent at the time of the enclosure, early became a centre of settlement, possibly due to readily available water supplies, proximity to the demesne, the manorial mill and some of the common fields.

From among the many examples in the local records we may note (xiv) that in 1561 Richard Lutley sold to William Hunt, *inter alia,* a cottage and meadow in Hateley Field, or that in the court of 1597 (xv) Thomas Cotton proved his claim to all the land formerly held by Richard Corston at Hateley. The only references to Spon Heath that we have are (a) that in 1633 (xvi) John Simcoxe died seized of land near to 'Spone' Heath, and (b) that in 1709 (xvii) John Shelton, Lord of the Manor, sold to Thomas Dudley 54 acres of 'Spann' Heath. We may note that within the 1,760 acres of the Manor Estate sold in 1820, there were 18 enclosures which included 'moor' as part of a field name, while there were many gorsty or broomy closes or leasows in the many conveyances of which we have record. The same record of sale lists two osier beds associated with mill dams.

Bearing in mind the localities of the types of vegetation we have mentioned and by referring to the chapter on vegetation by Rees and Skelding in 'Birmingham and its Regional Setting' and to Pitt's statement (xviii) on the subject we may now attempt to draw our picture.

The Tame, although mostly flowing through the forest, at times emerged to be bordered by a wide flood plain up to half a mile in width. Here the river, bordered by willows and alders, meandered through the swampy stretches thick with sedges, rushes and meadowsweet. Above the valley floor, the slopes were covered with a woodland of oak (quercus robur) with a little holly. Frequently the close canopy precluded much ground vegetation, but where it was more open, hazels and brambles were to be found as undershrubs. A ground vegetation of grasses such as wavy hair grass (Deschampsia Flexuosa) and clumps of soft grass (Holcus Mollis) together with such woodland plants as the barren strawberry (Potentilla) would probably occur.

The higher up the slope the more open the woodland would be and dense patches of bracken would then appear. By the time the summit of the hills was reached an open heath vegetation would predominate. There would be a scattering of scrub oaks with some hazel but the main feature would be the predominance of gorse and broom with dog-rose and blackberry brambles. Any ground flora would be the wavy hair grass with possibly the purple moor grass (Molinia) in the damper places. It is possible that in places bracken formed a dense cover to the exclusion of other plants. On the west facing slopes the heath appears to have been the calluna type with the typical flora of common heather or ling and possibly the vaccinias (whortleberry and bilberry). The damper conditions of Hateley Heath approaching marsh close to the Hobnail Brook would provide for a vegetation largely of the ericas, particularly the crossleaved heather, and vaccinias. The wavy hair grass would be replaced by the tufted variety and by the purple moor grass (Molinia) with probably rushes (Juncus) in the wetter parts.

Though most of the original vegetation has long since disappeared yet sufficient traces still remain to make the above picture at least a probability.

(i) W.B.M.C.R., 17 Jas. I., P.R.O., transc. Lissimore, W.B.R.L.
(ii) J. Reeves, *History of West Bromwich* (1836), p. 52.
(iii) W.B. Enclosure Act, 1802, transc. E. Lissimore, W.B.R.L.
(iv) R. H. Kinvig in *Birmingham and its Regional Setting* (1950), p. 121.
(v) M. Willett, *History of West Bromwich* (1882), p. 145.
(vi) M. Willett, op. cit., pp. 203-207.
(vii) M. Willett, op. cit., p. 205.
(viii) W.B.M.C.R., 16 Eliz., P.R.O., transc. Lissimore, W.B.R.L.
(ix) W. Pitt, *Agricultural Survey of Staffordshire* (1796), p. 97.
(x) Enclosure Act; see (iii) above.
(xi) M. Willett, op. cit., p. 219.
(xii) W.B.M.C.R., 7 Jas. I., P.R.O., transc. Lissimore, W.B.R.L.
(xiii) Do., 1 Jas. II.
(xiv) M. Willett, op. cit., p. 215.
(xv) W.B.M.C.R., 39 Eliz., P.R.O., transc. Lissimore, W.B.R.L.
(xvi) Do., 9 Chas. I.
(xvii) M. Willett, op. cit., p. 203.
(xviii) W. Pitt, op. cit., pp. 106-107.

The Early Days

OF the pre-Conquest history of West Bromwich, we know practically nothing. There are no traces of pre-historic or Roman occupation. Of Saxon times there remains but the faintest evidence. The name Bromwic itself, which was used for this area at the time of the Conquest, would appear to be acceptable proof of Saxon settlement. This would fit in with the pattern of Saxon name endings in this part of South Staffordshire. In addition to the '-wic' termination which we have in Smedeuwic (Smethwick) and Blockeswic (Bloxwich), there are several localities ending in '-ton', e.g., Tybinton (Tipton), Billestune (Bilston), and '-hale' as in Wirenhale (Willenhall), Rischale (Rushall) and Peleshale (Pelsall). It is probable that Walsall was of the same period. Then there are examples of the Saxon clearance terminations '-ley' and '-field' as in Sedgley and Wednesfield. However, we know nothing more detailed of West Bromwich than that 'Bructuin held it' (i), presumably at the time of Edward the Confessor.

It has been suggested that the name 'Bustleholme' is indicative of Danish occupation, but as the Scandinavian terminations of '-holme', '-by' and '-thorpe' are conspicuously lacking in this area, and as there is no proof of settlement at Bustleholme in Norman times and the name does not appear in documents before the 16th century, it would seem improbable, despite Danish incursion into the Midlands in the early 11th century.

We are on safer ground when we examine that invaluable record the Domesday Book which gives us so much information on the area as it was at the end of the 11th century. Here, under the entries for the Hundred of Offlow in the County of Staffordshire we find entries for Wednesbury, Bilston, Sedgley, Tipton, Bradley, Willenhall, Bescot, Smethwick, Handsworth, Barr, Perry and Harborne. These together with the entries for Dudley and Wolverhampton in Clent and Seisdon Hundreds respectively, almost completely cover the drainage area of the Upper Tame. The notable exceptions are those of Walsall and West Bromwich. However, the latter omission can be rectified, for we find in Northamptonshire:—

> Radolphus ten.de Willi.iij hid.in Bromwic; ter.e.iij car. In dmo.est una et x vil.et iij bord hbnt.iij car. Silva vij leuua longa et dim.leuua lata: Val.et val.xl s. Bructuin te.

This may be roughly translated as:—

> 'Radolphus holds from William FitzAnsculph) 3 hides in Bromwich; There are 3 carucates of ploughland. In the demesne is 1 carucate and the 10 villeins and 3 bordars have 3 carucates. The woodland extends for 7 leagues in length and ½ a league in width. The whole is valued at 40s. as it was formerly, when Bructuin held it.'

Of Domesday holdings, the king held Bilston, Wednesbury, Willenhall, Bescot, Bloxwich and Shelfield, later to become part of the royal Chase of Cannock; the Bishop of Chester held a number of berewicks or small hamlets, including Tipton and Smethwick, and the Canons of Handone (Wolverhampton) owned Wednesfield, part of Willenhall and Pelsall in addition to Wolverhampton possessions. The largest subject landowner however was one William FitzAnsculph, who from his main seat at Dudley could survey his holdings of Dudley, Sedgley, Bradley, West Bromwich, Aldridge, Handsworth, Barr and Perry. In addition to his South Staffordshire properties he held lands in other parts of the country including a number in Northamptonshire. Due to some unexplained confusion, Bromwic appears with his holdings in this county.

If we omit Wolverhampton and Dudley, West Bromwich appears to have had an importance that was comparable with the other locations referred to above with the exception of Sedgley. West Bromwich for taxation purposes was assessed as three hides, as were Wednesbury, Barr and Perry, compared with a total of eight hides at Sedgley. Of actual values quoted West Bromwich is given as 40/-, Barr as 5/-, Handsworth as 20/-, Rushall as 10/-. Bilston had increased in value from 20/- in Edward's time to 30/-, while in Darlaston the value had fallen from 30/- to 20/- over the same period. These values are overshadowed by that of Sedgley which was £10 in Edward's reign and it 'remains the same'. Thus West Bromwich was one of the more valuable properties of this district, though insignificant when compared with more prosperous areas.

When we examine the numbers of inhabitants we find the same sort of pattern. In West Bromwich there were no actual serfs but there were ten villeins, each holding his cottage and allotment of land in return for services, probably mainly day-work in the demesne fields, rendered to their Manorial Lord, Radulphus, vassal of FitzAnsculph. There were also three bordars, of lower status than the villeins, who probably held little more than their cottage in return for their services. This gives a total of fourteen households. If we accept Russell's estimation, quoted by Kinvig (ii) of an average of 3.5 persons per household we arrive at a Domesday population for West Bromwich of 50. This figure compares with populations of 90 for Wednesbury, 42 for Bilston, 40 for Handsworth and 32 for Rushall, if we calculate on the same basis. Sedgley, by a similar reckoning would have had a population of 210, including one priest and three serfs.

Populations such as these suggest that there was a greater density here than the average Staffordshire figure of 2 to 4 per square mile given by Darby and quoted by Kinvig (iii). The density for West Bromwich would have been about 5 per square mile. The sparse population may have been due to the natural lack of attraction for settlement shown by the area or to the historical results of its geographical location. This part of the Upper Tame valley appears to have been a 'no-man's-land' between the more favourable Trent, Avon and mid-Severn valleys and the rival Angle and later Saxon tribes occupying them. From the mid-9th century Mercia was subject to Danish raids which doubtlessly left wasted and depopulated lands behind them. Even after the Treaty of Wedmore in 878, the area was too close to the Danelaw for a guaranteed peaceful existence.

There is some evidence, not well substantiated, that Dudley and Wednesbury were sites chosen for two of a chain of Saxon fortresses constructed by Ethelfleda, widow of Ethelred, Earl of Mercia, about 916 A.D. A further Danish raid took place in 1016. Although we have no details, it is possible that this area received something of the 'pacification treatment' meted out in the north by the Norman Conquerors, for six years later we find reference to former settlements lying waste in South Staffordshire. Such were Bescot, Shelfield and Pelsall. Such were the conditions of Domesday South Staffordshire of which Bromwic formed a part.

(i) Domesday Book; quoted M. Willett, op. cit., p. 3.
(ii) R. H. Kinvig in *Birmingham and its Regional Setting* (1950), p. 119.
(iii) R. H. Kinvig, idem.

Mediaeval West Bromwich

BY the end of the 11th century, West Bromwich had settled down to a peaceful existence that it was to enjoy for the next five hundred years, apart from a couple of alarms, when a Poll Tax was levied to support the military activities in 1380, and again when a Muster Roll was taken in 1539 preparatory to a possible military conscription (i). Apart from these it was to remain a quiet backwater unaffected, so far as we can tell, by the alarums and strifes that disturbed more envied parts of the country. Slowly the woods were to recede and give way to the expanding agriculture and the population was to show a slow steady increase, little augmented by an immigrant contribution from beyond its borders.

At the commencement of this period, as we have already seen, a large part of West Bromwich was wooded. The extent is given as 7 leagues by $\frac{1}{2}$ a league (ii), which would be the equivalent of $10\frac{1}{2}$ miles by $\frac{3}{4}$ mile, taking the Saxon leuua to be 12 furlongs (iii). These measurements present some difficulty, as we cannot exactly locate the position in which they were taken. However, if we take the longer measurement as having been taken a little within the boundary and parallel to it, we find the estimation quite reasonable. This is supported by our previous assumption of the location of the woodland. It then might well follow that the smaller dimension is an estimate of the average width from the boundary up the slope towards the barer hill tops. This would give a forested area of 4,000 to 4,500 acres or about 80% of the parish.

The nucleus of agricultural West Bromwich was already established. Though the details given in Domesday are not exact, it appears that there were 7 carucates of arable land, one of which was within the demesne. Though the carucate was a measurement which varied with the locality, we may assume that, in an area of this type of soil, 80 acres per carucate is a reasonable estimation. This would give an arable area of about 560 acres. There is no mention of pasture. We cannot be absolutely certain of the location of the hamlet and its clearance but it is probable that it was centred on the site of the present Manor and Hall Green. Here is a position on a west facing side of a broad gently sloping

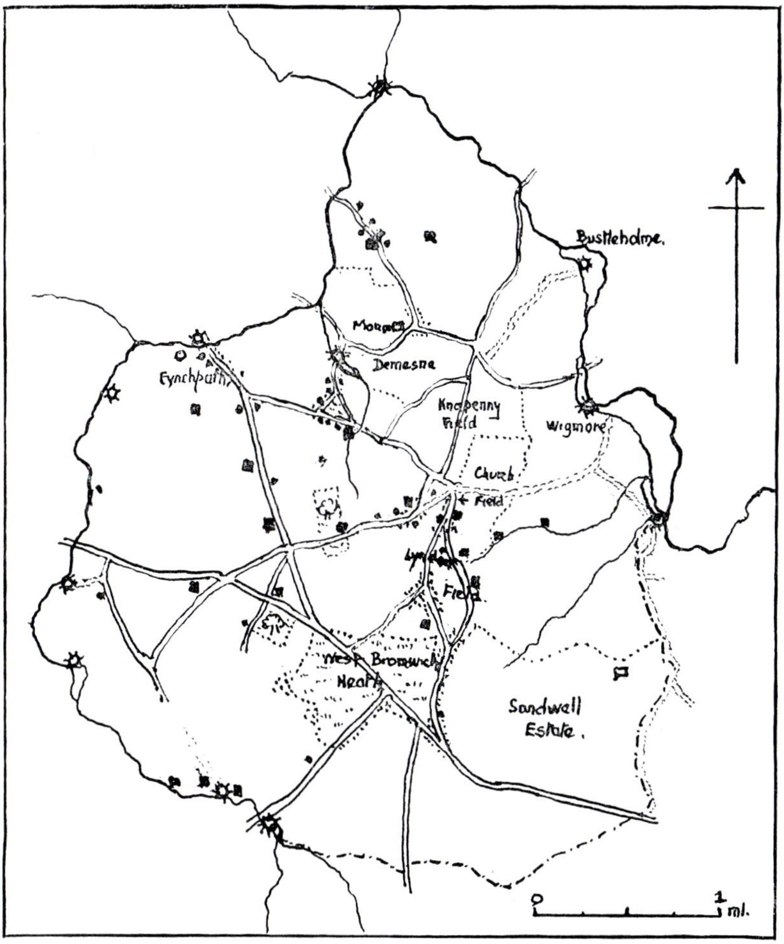
Bustleholme.
Demesne
Knapenny
Field
Wigmore
Church
Field
Field
West Bromwich
Heath
Sandwell
Estate.
0
1
ml.

valley of the Hobnail brook. The bottom of the valley was already clear of forest with its area of heath, the ancestor of Hateley Heath, and its small area of marshland. The forest to the north and east offered protection from the coldest winds. The water-table was within ten or twenty feet of the surface, so that while surface brooklets probably supplied the earliest needs, it was not difficult at a later date to sink shallow wells. If the forest margin was pushed back to provide for the 560 acres of plough-land, we find that we have the area which at a much later date was the Manorial holding and part of the open common field system of West Bromwich. That there is no record of pasture is hardly surprising on this site, though it would have been had the chosen position been in the Tame valley itself.

The fact that there is so little in Domesday about West Bromwich is sufficient to give a fairly clear picture of life here at the time. There is nothing beyond the mention of the arable land. It is a pity that we have no reference to animals. It is pretty certain that horses, cattle, sheep and pigs were all kept by the few inhabitants. They may have used horses or oxen for the ploughing. The absence of meadow is not an insuperable obstacle, for the heath-land certainly provided a substitute, albeit a poor one, as it did up to the time of the final enclosures (iv). The open woodlands undoubtedly provided grazing, for though not considered worthy of recording in West Bromwich, it was recorded at Barr and at Rushall.

Similarly a Deed of 1197 (v) granted to the Burgesses of Walsall right of pannage at acorn time. The Manor was insignificant. The Lord of the Manor had no serfs, retainers or servants. The population was not large enough to warrant a church or a priest. There was no corn mill. Either the corn was ground by hand or taken out of the parish for grinding. The nearest mills, both on the Tame, were the Wednesbury Mill, about a mile away, and the Handsworth Mill, about two miles distant. The cottages were probably wattle and daub with thatched roof, as stone was not readily obtainable. It appears to have been a proverty-stricken community, mainly self-supporting but with little to attract outsiders.

The changes which took place during the next 500 years were those of a very slow growth of population, reduction of forest area, growth of land under cultivation, alienation of Manorial property rather than a change of pattern of life and conditions of the people. These topics will be dealt with separately.

The biggest change occurred a hundred years after Domesday. About the year 1175, William, son of Guy de Opheni, gave the area of Sandwell Park, then forested, to found a Priory. This was confirmed in 1180 by Gervase Paganel, Baron of Dudley, overlord of the Opheni (vi). This gift included an unspecified area around Sandwell and an earlier hermitage, lying partly in West Bromwich and partly in Handsworth. Thickly wooded, this land was without settlement though two assarts and a solum are mentioned. The gift carried with it the right to construct mills, of hunting, assize of bread and beer, of gathering litter for the kitchen, of firewood and of timber for building repairs; in other words the rights pertaining to a forested area. There was an additional privilege of grazing animals of any sort anywhere within the manor, i.e., on the lord's wastes and in his woods. Two other interesting items are included in the deed, a well at Wich and a mill at Grete (Great Bridge). This was probably on or near the site of the later Sheepwash Mill, the earliest recorded mill that we have within the parish. It was situated almost as far as possible from the Manorial settlement and it would seem probable that another much closer to the demesne would have been constructed by this time. The first hundred years of this Benedictine Monastery, a subsidiary of the Abbey of Hales in the diocese of Worcester, saw the same slow agricultural development typical of West Bromwich. The Taxation of Pope Nicholas IV, 1291 (vii), shows the temporal possessions of the Prior of Sandwell as being three carucates, say 250 acres, of arable land, valued at 10/- per annum, pasture land to the value of 10/- per annum, rough grazing worth ½ marc, and two mills (presumably the one at Grete and another in the Sandwell region) with an annual value of 10/-. His most valuable asset was 40/- per annum in rents. This would represent a considerable hamlet. The location of this is a matter of conjecture but was probably in the region of Sandwell Green. Two years later was an eventful year for the Priory, for the Prior was in legal dispute with the Lords of the Manor over the grazing rights on the common arable fields (viii). Such disputes were a frequent occurrence through the centuries. In the same year a probationary monk, working as a tanner, fell into the brook and was drowned (ix).

Early in the 13th century another area was bestowed on the Abbey of Hales (Halesowen), as a donation when the Abbey was founded. This area was Friar Park.

We must now jump a period of another hundred years to 1380, when in the turbulent times of Richard II, a Poll Tax was levied. A copy of the return for West Bromwich which exists (x), contains more detail of the people than the Sudsidy Rolls of 1327 and 1330 (xi). This return reveals that little had changed here in two hundred years. There was slow expansion and progress. The number of adults was now 58 suggesting a total population of about 90. There is no record that the Black Death of thirty years earlier had reached the district and so reduced the numbers. There are seven families whose occupation is not stated, two of whom, the de Marnhams and the Bassetts, at that time each held a moiety of the Manor, though the latter is not set down as the Lord of the Manor. Of the others, eleven are stated to be agricultural labourers, seven cottagers and five workmen. The labourers I take to be the equivalent of the earlier villeins, and the cottagers, who appear as people of less wealth, to be the cottars. The workmen were probably craftsmen in their own right. In addition, mention is made of a blacksmith and a basket maker.

One change had come about by this time. The population was no longer living in one small hamlet close to the Manor at Hall Green, but was scattered throughout the parish. Mention is made of Luttylhay, Sponne, Lyndon, Ruggiacre and the 'Hole'. Thus the seeds were sown of the later settlement pattern. The Statute of Merton in 1236 allowed a Lord of the Manor to enclose the common land or waste provided sufficient was left to meet the needs of the enfeoffed tenants. Enclosure in the wastes of West Bromwich had commenced at least as early as the beginning of the 14th century, for we have records of Deeds of Sale or Lease of land (xii) either by the Lord of the Manor himself or resales with his sanction. These, prior to 1346, were all in the Finchpath (Hill Top) to Wednesbury Bridge area, but in 1348 (xiii) one John Sarsone conveyed to Richard Lovekyns six selions of land in one of the open common fields. By this date, here at least, the strips in the comomn fields were no longer freshly allocated each year but had become the property of an individual and as such capable of transference, as any other plot of ground. Such transactions are frequently recorded from this time until the final enclosure in 1802.

Thus, until the 16th century the bucolic life of a rural West Bromwich went its uneventful way, little affected by national troubles and only occasionally disturbed by quarrels and felonies, records of which are to be found in the Coram Rege Rolls and the Stafford Assize Rolls. Slowly the population increased to about 250 at the time of the Muster Roll of 1539 (xiv) and spread itself out into the half dozen or so hamlets. They were the agricultural workers and the craftsmen and traders. There were the woodsmen, the tailor, the cook, the brewer and the thatcher. There was Richard the cleric and William and Symon the millers (xv), John Symcocks the blacksmith and Robert Bonde the basket maker (xvi), and we must not omit Agnes the weaver (xvii).

The first church was built about 1125. The Manor house dates from about the 14th century. Clearing and enclosure had extended to all parts of the parish, parts of the Manorial wastes being enclosed at Grete Green and Black Lake in 1573 (xviii). More mills had been built, Bromford Mill before 1272 (xix) and Turton's Mill and Bustleholme certainly before the end of the 16th century. But as yet there was little or no sign of the industrialisation that was to come.

(i) Muster Roll, 1539, *Staffordshire Historical Collection,* Vol. IV (New Series), p. 245.
(ii) Domesday Book.
(iii) R. H. Kinvig, *Birmingham and its Regional Setting* (1950), p. 121.
(iv) W.B. Enclosure Act (1802), transc. Lissimore, W.B.R.L.
(v) F. W. Willmore, *History of Walsall* (1887), p. 159.
(vi) M. Willett, *History of West Bromwich* (1882), pp. 145-146.
(vii) S. Shaw, *Natural History of Staffordshire* (1798), I.
(viii) Stafford Assize Roll, 21 Ed. I., S.H.C., Vol. VI, (i), p. 236.
(ix) F. W. Hackwood, *History of West Bromwich* (1896).
(x) *Staffordshire Historical Collection,* Vol. XVII, p. 174.
(xi) *Staffordshire Historical Collection,* W.S.L., Vol. VII, p. 229 & Vol. X.
(xii) M. Willett, op. cit., pp. 201-224.
(xiii) M. Willett, op. cit., p. 205.
(xiv) H. H. Prince, *Old West Bromwich* (1926), pp. 46-47.
(xv) M. Willett, op. cit., p. 203.
(xvi) H. H. Prince, op. cit., p. 42.
(xvii) Stafford Assize Roll, S.H.C., Vol. VI, (i), p. 268.
(xviii) W.B.M.C.R., 16 Eliz., P.R.O., transc. Lissimore, W.B.R.L.
(xix) F. W. Hackwood, *Oldbury and Round About* (1915), p. 45.

The Fields of West Bromwich

As we have seen in a previous chapter, the field system at the time of Domesday consisted of arable land of which some eighty to a hundred acres lay within the demesne in the Hall Green area to the ENE side of the Hobnail valley. The rest, some 480 acres, most probably lay around the south-east and southern gently sloping sides of the same valley. As was usual with the Saxon Field system, it is probable that two-thirds would be under cultivation at any one time but we have no trace of boundaries dividing the lands of those days into three parts. It is not improbable, however, that the building of the church about 1125 A.D. and a small settlement around it, indicates the limits to which the cultivated land extended at that time. The area of the field would be extended as the population grew and as the forest bounds receded.

We have no records before the mid-14th century of the naming of individual fields. When the Abbot of Hales brought the action (i) in 1260, he was claiming right of common pasture over 70 acres of arable land, but it did not stipulate in which field. The first reference to a field by name is in 1348 (ii) when there is a record of the sale of six selions in Wygmorefield. It is unlikely that this was one of the original fields, for although we have no exact knowledge of its position or extent, it was probably near to the Wigmore of today (022936). Willett (iii) states that Wygmorefield lay between Churchfield and the Newton boundary, i.e., the same area, though without substantiating the statement. It appears that at times Wigmore Field was an extension of Church Field while at other times a separate entity. The first mention of the Church Field by name was in 1637 (iv), again a conveyance, this time of four selions in Church Field and four selions in Styecroft. This suggests a fourth field, i.e., Styecroft, further from the original centre. The oldest of the fields appears to have been Napeney Field, which was first mentioned by name in 1516, when it was referred to as the destination of a lane. This field name appears under various spellings. In 1516 it was Guarpanell Field, in 1709 Knappenhill Field, and by the time of the enclosure in 1802 it had become Napney Field.

The fields were not only extended but they were sub-divided at times. The Napney Field and Church Field, the two earliest fields, maintained their position more or less, though changing in area, up to the time of the enclosure. The Wigmore Field, which had such an early reference does not survive beyond the beginning of the 18th century, at about which time that area passed into private ownership. The area to the south and south-east of the church became one of the common fields early in the 17th century. I am not certain if this was all considered as one field or no, for though most of it was known as Lyndon Field from 1633 (v) until 1802, Stye Croft, which was the south-east part of it, is separately named from 1637 to 1802. In the details of the enclosure, the northern part of this field is named as Windmill Field. There are two other names used, Liddiate or Lydiate Field, which may well have been an alternative name for part of Napney Field, and a Hately Field, of which only one reference (vi) survives.

There are no contemporary maps showing the location of the fields, but we have two documents which are useful in this respect. There was drawn up in 1720 a ' Survey of Lands in the Mannor of West Bromwich ' (vii) which enumerated all the fields, open and enclosed, which at that time formed the manorial holding. Each field was marked with a grid reference which was intended to correspond with a grid on a map, now unfortunately lost, but which it is possible to reconstruct with some degree of accuracy. The other document is the detail of the Enclosure Act, which in addition to having an accompanying map, does indicate the position of each holding by quoting its bounding lane, footpath or adjacent properties. From these two records it is possible to give a fairly accurate estimate of the position of the Napney and Church Fields and to a less extent, the Lyndon Field.

When we look at the problem of the extent of the open fields, there are several factors which we must bear in mind; first the growth of the fields to meet the growing demand; secondly, the vesting of the ownership of the strips of ground in one individual, leading to efforts of consolidation of the strips into larger holdings and finally the actual enclosure of such holdings and their loss to the open fields. The first point has already been considered. The second has already been referred to above and quoted by the examples of 1340 (viii) and a sale of 1670 (ix).

An examination of the above quoted Survey of 1720 (x) shows several examples of enclosed ground on the margins of the open fields, having been named as the ' New Inclosure '. We have few references to actual extent. We know that at Domesday the arable area outside the demesne was six carucates which we estimated as about 480 acres. From that time on until the end of the 16th century, our only material source is contained in numerous conveyances. The writers of these for the most part were content with defining the ground as 'late Crompes Ground' or 'in Fynches-path by the way from Wednesbury to Birmingham'. Areas, if mentioned at all, were usually in carucates, or later, in 'days work' or selions. The latter was approximately $\frac{1}{4}$ acre and the 'days work' while variable was, here in West Bromwich, roughly one acre in extent.

The Rent Rolls, from which we might expect to obtain information are not too helpful in this matter, for they are usually confined to the names of the tenants and amounts of the rents. The Roll of 1649 (xi) is a little more detailed, though even here the entry is usually for 'one tenement and certeyn lands' in the case of the smaller holdings. In the case of William Osborne, however, he was paying rent for two fields with '1 acre of land in ye comon feild'. By 1720 (xii) the rented lands in the open fields had been reduced to 70 acres, which with 49$\frac{1}{4}$ acres of 'new Inclosures' gives a total of 119$\frac{1}{4}$ acres. By 1802, the total area was about 95$\frac{1}{2}$ acres.

The deforestation and the enclosure of the land in the further parts of the parish had started at least by the early 14th century. The numerous Deeds of Conveyance quoted by Willett (xiii) and which unfortunately cannot now be traced, tell of the enclosure and transactions in land and houses over a period of four hundred years.

From these it is possible to trace the pattern of development. The area of Fynchpath (Hill Top) and Wednesbury Bridge is the first to appear with numerous records from about 1320; Ruggiacre (Ridgeacre) is mentioned in 1346 and Goldscroft (Hall End) in 1382. The Hill Top area had extended to Black Lake before 1446. Despite the existence of a mill at Grete in 1180, already noted we do not find records of land transactions along the western part of the Tame valley until the 16th century; Great Bridge in 1521 and the Turton's purchase of two mills and one meadow in 1583.

These conveyances were in some cases sales of freehold outright, or under a quit claim or chief rent, or leases or rents of various types. While rentals of manorial lands in the older part of the parish were usually related to isolated and scattered enclosures, probably held by cottagers in the hamlets, those further afield were often grouped as farms with the appropriate tenement and appurtenances.

In 1720 there were twenty such farms or smallholdings each with over 25 acres, while in 1649 (xiv) eighteen such were included. It is interesting to note that though the manorial lands were finally disposed of in 1820, and though industrial development made such rapid strides in the 19th century yet 17 farmers were listed in 1818 (xv) and 18 in 1855 (xvi). In 1969 six still struggled for existence.

One of the serious problems created by the grazing of animals on the wastes was the necessity of fencing the arable against the depredations of such animals on the crops. The responsibility in the case of the open fields was that of the holder of the strip adjacent to the ring fence. A break in one man's part of the fence was likely to lead to a loss of crops by others. The usual method of fencing here was by the planting of holly or hawthorn hedges. Holly still is to be found on the site of what was part of the boundary of the Church Field and there are several references to the use of hawthorn.

In 1728 (xvii) Thomas Jesson rented two pieces of ground to Samuel Lloyd but agreed to 'quick-set the gaps'. The results of breaks occurring in the fences are all too often recorded as in the following: 'Michael Pountney fined 12d. for breach of his hedge between the followe field and the oat field, through which the beasts of divers persons there pasturing in common have passed into the cornfield there to the injury of his neighbours' (xviii). Regulations on this matter were laid down from time to time; 'a day is given . . . and similarly to all others who have not made their mounds and hedges in the same lane'. This was in 1573 (xix).

The enclosure of the wastes led to another problem. In 1606 Nicholas Drurie had stopped the ancient common footpath on his lands leading towards Greete Bridge. He was warned and told to build a stile (xx).

Much of the enclosure of the wastes and commons was quite unauthorised and was done by squatter settlement in the 17th and 18th centuries. Such were for the most part in the meanest type of building. They were described as 'little better than a temporary hut wattled with sticks and plastered in some cases with mud and covered with branches of trees, straw or turf' (xxi). Some were still to be seen in 1836 at Old End and Cutler's End, on the margins of the Heath. There was no attempt made to remove them so far as can be ascertained. The occupants were usually fined about 4d. in the Manorial Courts. There were no less than 71 such cases in 1685 (xxii). A footnote to the Rent Roll of 1630 reads 'Cotages upon the waste. Itm. the rent payed for all the rest of the cotages upon the waste and for some lytle parcells of wast grounds which are not much to be improved because they are poor people that hold them is 10s. 10d.'

(i) Stafford Assize Roll, 21 Ed. I., S.H.C., Vol. VI, (i), p. 223.
(ii) M. Willett, *History of West Bromwich* (1882), p. 205.
(iii) M. Willett, op. cit., p. 168.
(iv) M. Willett, op. cit., p. 224.
(v) M. Willett, op. cit., p. 106.
(vi) M. Willett, op. cit., p. 218.
(vii) P.R.O.; photostat W.B.R.L.
(viii) M. Willett, op. cit., p. 205.
(ix) M. Willett, op. cit., p. 220.
(x) P.R.O.; photostat W.B.R.L.
(xi) W.B. Rent Rolls, P.R.O.; photostat W.B.R.L.
(xii) W.B. Manorial Survey, P.R.O.; photostat W.B.R.L.
(xiii) M. Willett, op. cit., pp. 164-224.
(xiv) W.B. Rent Rolls, P.R.O.; photostat W.B.R.L.
(xv) Parsons & Bradshaw, *Birmingham and District Directory,* 1818.
(xvi) White's, *Birmingham and District Directory,* 1855.
(xvii) M. Willett, op. cit., p. 225.
(xviii) W.B. Manorial Court Rolls, 15 Jas. I., P.R.O., transc. Lissimore, W.B.R.L.
(xix) W.B.M.C.R., 16 Eliz. I., as above.
(xx) W.B.M.C.R., 4 Jas. I., as above.
(xxi) J. Reeves, *History of West Bromwich* (1836), p. 56.
(xxii) W.B.M.C.R., 1 Jas. II., as above.

West Bromwich Agriculture

AT Domesday and for long after, the customary arable procedure was to divide the land into three parts and to rotate their usage. In any one year, one field, the ' Winter Field ', would be under, say, wheat; the second, the ' Spring ' or ' Lenten Field ', would be in spring sown crop such as oats, while the third field would lie fallow. The fallow field was thrown open to common grazing all the year. This served the dual purpose of providing a change of feed (even if a poor one) from the heather, bracken or gorse-ridden heaths, and at the same time a manuring of the ground. This practice has already been quoted above (i).

Pitt (ii), writing of South Staffordshire agriculture in 1817, stated that the gravelly loam of the higher ground was best suited to the cultivation of turnips and barley, while the lower ground, cool and moist, was better suited to wheat, oats, or meadow and pastures. Reeves (iii), in 1836, stated that the area produced good crops of wheat, oats and grass. There is little reference to the crops actually grown in the West Bromwich open fields. In 1606 (iv) we find reference to the winter or rye field and the lenten or oat field. These would probably have proved more reliable crops on this soil than wheat and barley. However, by 1685 (v), we have an order that ' horses and cattle may not be turned into any common field before all the wheat and crops have been carried '. Although in West Bromwich we have no indication of the order of rotation, the Rev. Ames (vi) in his comments on daily life in Bilston, tells us what was happening not far away. The rotation was as follows:

	Colepit Field	**Windmill Field**	**Middle Field**
1724	Fallow	Barley	Hard corn (wheat)
1725	Hard corn	Fallow	Barley
1726	Barley	Hard corn	Fallow

Thus the wheat crop always followed the fallow. It is probable that the West Bromwich rotation was rye (or wheat) followed by oats and then fallow. It was the custom to harvest all the strips in the one field at the same time so far as possible, and then to turn the cattle into the field. There is plenty of evidence in the Manorial Court Rolls that some people were over-anxious and apt to turn in the cattle before all the corn was harvested. This led to a number of fines being levied. The practice of fallowing the ground continued even after the enclosing of land. The deed of 1728 (vii), already referred to, by which Thomas Jesson let two pieces of ground to Samuel Lloyd, stipulated that only one of them should be ploughed in any one year. Even as late as 1846, a lease of the Bustleholme grounds contained the restriction that not more than three grain crops might be taken in successive years, nor that more than eleven of the 72 acres might be under arable crops during any of the last three years of the lease.

We know little of other crops grown. In 1533 an Act was passed which ordered that 'every person having in his occupation three score acres of land, apt for tillage, shall sow one rood with linseed, otherwise called flax or hemp seed and also one rood for every forty acres more'. This cannot have affected many in this area of mainly smallholders. It is noticeable however that several of the poorer citizens had an 'hempleck' attached to their cottage (viii). It is possible that this provided something of a livelihood for them.

In 1606 (ix), it was ordered that 'Anne James, widow, shall not from henceforth burn hemp-stalks or sheaves in her backyard to the annoyance of her neighbours'.

It is certain that, in West Bromwich, flax growing never assumed the importance which it did in Bilston, where the names of 19 flax dressers occur in the parish register between 1720 and 1730 (x). At a later date the growing of flax was discouraged, probably because it tended to impoverish the soil. This was usually done by demanding an increased rent if the crop should be grown, e.g., the rent of the Bustleholme Mill estate was to be increased by £20 per acre for 'planting flax, hemp, mustard or woad' (xi). It would seem that restriction on the farming methods of tenants was common practice in the 19th century.

The most complete example we have is contained in a 21-year lease of a farm in 1805 (xii) in which the lessee agreed to pay extra rent for each acre broken for hemp, flax or rape seed without permission or during the last three years of the lease; not to burn off more than 27 acres; not to plant corn on two successive years without fallowing or sowing grass; not to plough more than could be manured; to return to the land any hay or straw dung arising from the hay or straw grown there; to supply and spread one load of manure in the last year for each load of hay or straw taken off the fields; to summer fallow, in each of the last three years, nine acres of arable and to lay down nine acres with 12 lbs. of clover and two bushels of best grass seed per acre; to autumn graze by sheep only and to allow no horse or cattle on the ground after 21st December in the last year and never to have more than 27 acres of all grains during the last three years.

Though horses, cattle, sheep and pigs were all kept, there is little specific evidence on local animal husbandry. The documented references are usually the casual and incidental. The practice was to turn the animals on to the 'wastes', i.e., the heaths and open woodlands. There the burning of gorse and ling would in time reduce the amount of this type of vegetation and encourage the growth of bracken, while heavy grazing, particularly by sheep, would increase the areas under wavy hair grass and fescue. At the earliest possible moment after harvest, the common arable fields were thrown open and the animals moved to them. This would give a beneficial bite to the beasts, which would effectively glean any grain left; they would clean up most of the weeds and manure the ground.

Reference has been made in an earlier chapter to efforts made to keep the animals out until the harvest was declared cleared. There was probably at times a tendency to overstock the wastes, to the detriment of all concerned. Pitt (xiii), in 1796, referred to West Bromwich Heath, among others, as being a sheep-walk and warren. Again (xiv) he stated that about 11,000 sheep were summer grazed on about 6,500 acres of Sutton Park. It is doubtful if West Bromwich Heath could carry anything like that density with the cattle, etc., in addition. In fact the use of the common grazing ground was well regulated. In 1606 (xv), it was laid down that no man should keep more than one beast or five sheep on the common for every acre of ground he occupied. In 1633 (xvi), John Granger was fined 3s. 4d. because he 'surcharged the common with his beasts'. In 1688 (xvii) four men were each fined 1s. for keeping and pasturing on the commons within the Manor the sheep of foreigners. It was ordered that anyone so offending again would pay 39s. 11d.

The keeping of pigs in the open has always been a problem, since it is the nature of the beast that it will root up the turf. We have already (xviii) noted that in Walsall pannage was permitted at acorn time. Presumably, at such a time the appetite would be satisfied with acorns and there would be less rooting. In West Bromwich some pigs were kept in sties, but those ranging loose had to be ringed in the nose (xix). In 1609 (xx) Thomas Dudley was fined 20d. for keeping an unringed swine.

The nature of dogs does not change, for in 1685 (xxi) it was ordered that 'any cottager living within the lord's waste, who keeps a dog which injures or chases the cattle on the common pasture shall have it put away'. It does not say what would be the penalty for chasing rabbits, though the heath was recognised as a warren, the prerogative of the lord of the manor. We have already mentioned Pitt's reference to it as a warren in 1796. The right of Free Warren was granted to the Lord of the Manor in 1293 as the result of an inquiry (xxii). This right was held until the Enclosure. A house, the lodge (near to the present Lodge Road), the property of the Manor, was established on the heath for the warrener. In 1695 (xxiii) a William Smith paid £3 per annum as rent for the lodge and the warren. The lodge remained until about 1854.

One of the problems of extensive grazing over open common land was that the animals would go astray. It was necessary to establish a 'pound' and an officer to look after it. The traditional site of the pound, now greatly reduced in size, was close to the church. In 1685 (xxiv) Walter Wyley was sworn in as 'Pinner' or pinder. Any animal which remained unclaimed after a period in the pound became the property of the lord of the manor. In this way he acquired a mare and filly in 1609 (xxv) and a bullock, ewe and two lambs in 1617 (xxvi). It is not known when this privilege was assumed for 'right of Wayf' was denied the lord of the manor at the aforementioned Inquest of Quo Warranto. From time to time the pound was broken open and the animals were removed. This felony was regarded as serious. In 1689 (xxvii) William Tray was fined 30s. 4d. for this crime.

The woods remained the property of the manor. When they were much reduced in area, they were managed and cropped. Areas at Spon, Ridgacre and Cordley were retained as coppice timber. Tenants were granted right of grazing within the coppiced woods. In 1626 (xxviii) there were 122 acres of such woodland, 50 acres of which were 14 years old and the rest 6 or more years old. A footnote to the rent roll of 1695 states that the Sponn Coppice was cut in 1694 and the Caudley and Rudgiacres Coppices were both cut in 1695. The tenants would have the right of gathering firewood but no timber beyond that. In 1634 (xxix) four men were each fined 2d. for employing others to bind faggots in the lord's waste.

Reeves (xxx) stated in 1836 that 'of fruit not much is grown, but many colliers and workmen have their little gardens and some of them assiduously cultivate the gooseberry and currant'. Nevertheless, some fruit-growing had long been carried on. The Inquest of the Sandwell property (xxxi) in 1526 found, inter alia, 'a messuage with garden and orchard'. From 1720 to 1740 (xxxii) Sir Samuel Clarke did much to improve and extend the orchards of the manor, planting apples, pears, plums, cherries, grapes, peaches, apricots, nectarines and figs. With what success we know not.

(i) W.B.M.C.R., 15 Jas. I., P.R.O., transc. E. Lissimore, W.B.R.L.
(ii) W. Pitt, *History of Staffs.* (1817), p. 161.
(iii) J. Reeves, *History of West Bromwich* (1836), p. 95.

(iv) W.B.M.C.R., 4 Jas. I., P.R.O., transc. E. Lissimore, W.B.R.L.
(v) W.B.M.C.R., 1 Jas. II., P.R.O., transc. Lissimore, W.B.R.L.
(vi) Bilston Parish Register, 1684-1746, Parish Register Soc.
(vii) M. Willett, *History of West Bromwich* (1882), p. 225.
(viii) W.B. Rent Rolls, 1630, P.R.O.; photostat W.B.R.L.
(ix) W.B.M.C.R., 4 Jas. I., P.R.O., transc. Lissimore, W.B.R.L.
(x) Bilston Parish Register, 1684-1746, Parish Register Soc.
(xi) Smith Deed Collection, W.B.R.L.
(xii) Smith Deed Collection, W.B.R.L.
(xiii) W. Pitt, *Agricultural Survey of Staffs.* (1796), p. 120.
(xiv) Ditto, p. 140.
(xv) W.B.M.C.R., 4 Jas. I., P.R.O., transc. Lissimore, W.B.R.L.
(xvi) W.B.M.C.R., 9 Chas. I., P.R.O., transc. Lissimore, W.B.R.L.
(xvii) W.B.M.C.R., 4 Jas. II., P.R.O., transc. Lissimore, W.B.R.L.
(xviii) F. W. Willmore, *History of Walsall* (1887), p. 159.
(xix) W.B.M.C.R., 4 Jas. I., P.R.O., transc. Lissimore, W.B.R.L.
(xx) W.B.M.C.R., 7 Jas. I., P.R.O., transc. Lissimore, W.B.R.L.
(xxi) W.B.M.C.R., 1 Jas. II., P.R.O., transc. Lissimore, W.B.R.L.
(xxii) Plea of Quo Warranto, 21 Ed. I., S.H.C., Vol. VI (i), p. 248.
(xxiii) Rent Roll, 1695, P.R.O.; photostat W.B.R.L.
(xxiv) W.B.M.C.R., 1 Jas. II., P.R.O., transc. Lissimore, W.B.R.L.
(xxv) W.B.M.C.R., 7 Jas. I., P.R.O., transc. Lissimore, W.B.R.L.
(xxvi) W.B.M.C.R., 15 Jas. I., P.R.O., transc. Lissimore, W.B.R.L.
(xxvii) W.B.M.C.R., 1 W&M, P.R.O., transc. Lissimore, W.B.R.L.
(xxviii) Rent Roll, 1626, P.R.O.; photostat W.B.R.L.
(xxix) W.B.M.C.R., 10 Chas. I., P.R.O., transc. Lissimore, W.B.R.L.
(xxx) J. Reeves, op. cit., p. 95.
(xxxi) M. Willett op. cit., p. 154.
(xxxii) M. Willett, op. cit., appendix on fruit trees.

Land Tenure

LIKE the rest of the country, from Domesday, West Bromwich was subject to the Feudal System. Unlike most of the country to the north of the Tame, which was vested in the crown, this manor was held in ' knight's fee ' by the lord of the manor, from an immediate overlord, at that time Lord of Dudley. By this he had to swear fealty and pay homage to the overlord and if called upon, to give armed service, i.e., go to the aid of the overlord, suitably armed and accompanied by his armed retainers.

There were no serfs in the manor and the rest, freemen, held their land by socage, that is the rendering of services within the manor. With the passage of time most of the services were superseded by monetary payments and a system of rentals arose. By the 14th century the manorial wastes were being alienated by sale or lease (i) but the feudal system did not die easily. The essence of that system was that no man had absolute ownership of land, and after the decline of feudal days, the land tenure system survived largely because it was a valuable perquisite of an overlord. For that reason all conveyances had to be carried out publicly before witnesses of repute, a suitable fee being due to the manor. Thus, when John Golde of Fynchespath conveyed a messuage and land to Randolph, son of Alexander, c. 1320, the witnesses were no less than Richard Bassett of Bromwich (Lord of the Manor), Roger Illori (Lord of Bescot Manor), William de Delaston (Lord of the Manor of Darlaston) and Richard the Cleric (ii). For this reason attendance at the various manorial courts was obligatory, unless the person concerned ' essoined ' (excused) himself beforehand, and all matters were there made public.

All new occupiers of property still had to swear fealty, which included the undertaking to observe the customs and pay the dues. In 1580 (iii), Simon Ryder attended the Court Leete of the Lady of the Manor and compounded his dues of 1/10 of a knight's fee for his land. Thomas Huntback transferred Bustleholme Leasows to Francis Syncocke and acknowledged the transaction before the Manorial Court of 1597 (iv) and paid a fee of 7s.

By the 16th century the acknowledgment of services due to the manor by the holders of freehold was reduced to the annual payment of a small token sum, the ‘ Chief Rent ’. In 1630 (v) ‘ the wholl yeares Cheife rent is £10 11s. 10d.’

Another fee, heriot, was payable on the death of the holder of a property, whether leasehold or rented. This usually took the form of the best beast of the deceased and/or a sum of money. Before the court of 1597 (vi) notice was given that Richard Bate had died, seized of a messuage and land at Wystie End (Walsall Road). A heriot of 18d. and the best animal was levied. The rent roll of 1649 shows the heriot to be paid at the death of each lessee. In some cases the heriot could outlast the lease. In 1648 (vii) William Osborne took up a lease for 21 years by indenture, which carried the clause ‘ 30s. for a heriot if the Sd. Wm. dye within ye Sd. terme or if he lives then 30s. at the expiration of his life ’.

Another charge on the estate of the deceased was that of ‘ dower ’. By this a widow was entitled to claim one-third of the estate. In 1223, Geva Bassett, widow of the Lord of the Manor, sued for her dower of one-third of the value of a mill and meadow then in the possession of the Prior of Sandwell (viii).

Manorial land was held by the lessees by leases of definite or indefinite duration. Those of definite duration were for a fixed number of years, usually 21 but sometimes longer. In 1560 (ix) Henry Jesson leased Black Lake to George Lowe for a period of 99 years. The indefinite lease was based on a length of one or more lives, not exceeding three. Thus the lessee could obtain a lease for the duration of his life. In addition he could nominate one or two close relations in which case the lease would be of effect for the longest of the three lives, heriot being payable at each death. In 1626 (x) John Turton held land by indenture, for three lives, his own, aged 50, Aga, his wife, aged 50, and John, their son, aged 24. By 1695 (xi) many leases on such terms were converted to a 99 year period. The indefinite lease has not been legal since 1925. There were few leases for a period shorter than 21 years though one or two were for less than three years. Leases almost always referred to the larger properties, farms or areas of land. The less valuable holdings, cottages, hemplecks and pingles, were usually let at an annual rental and stated to be ‘ at will ’.

It was not unusual for a rent to be in kind instead of, or as well as, in legal coin. The rent payable for the lease of the lodge and warren frequently included an item for so many rabbits. In 1731 (xii) Henry Talbot was paying £4 and ‘ 52 couple of rabbets ’ per annum. At the Manorial Court of 1686 (xiii) it was reported that Humphrey Rogers had died seized of a tenement and lands in Grete held at an annual rent of 8s. and two capons, and a heriot.

Token rents were occasionally agreed. In 1667 (xiv) Edward Groves was granted the use of William Turton's well on an annual payment of one peppercorn if called on to do so. Other examples quote the payment of one rose on Midsummer Day and another of one grain of corn at Michaelmas.

In many leases it was the practice to include clauses reserving to the landlord rights of minerals, timber, etc. In contrast, on sale absolute such rights passed to the new owner. At one time, to emphasize that the new owner was entering on these rights, he performed certain symbolic acts before witnesses of credit. Simon Ryder (xv) records in 1597 witnessing a new occupant taking his knife and cutting a turf and also a 'twigge of a young oke standing in the old blade mill poole side'. By such acts the memories of the witnesses might be reinforced in a day when the safe storage of documents was problematical.

Where the tenancy of a property was held in return for non-free service or the more menial services, the names of the tenants and relevant details were inscribed in the manorial rolls. The tenant would then receive a copy of the entry which became his title deed. Such a property was termed 'copyhold'. In view of the lack of serf labour and the smallness of the manor and population, it is hardly surprising that few examples of copyhold existed. However, an Indenture of 1661 (xvi) refers to a property as being adjacent to the 'copyhold of the late William Lane'.

(i) M. Willett, *History of West Bromwich* (1882), pp. 164-224.
(ii) M. Willett, op. cit., p. 203.
(iii) Simon Ryder, *Hypomnema,* W.S.L.
(iv) W.B.M.C.R., 39 Eliz., P.R.O., transc. Lissimore, W.B.R.L.
(v) W.B. Rent Rolls, 1630, P.R.O.; photostat W.B.R.L.
(vi) W.B.M.C.R., 39 Eliz., P.R.O., transc. Lissimore, W.B.R.L.
(vii) W.B. Rent Roll, 1649, P.R.O.; photostat W.B.R.L.
(viii) F. W. Hackwood, *History of West Bromwich* (1895), p. 16.
(ix) M. Willett, op. cit., p. 215.
(x) W.B. Rent Roll, 1626, P.R.O.; photostat W.B.R.L.
(xi) W.B. Rent Roll, 1695, P.R.O.; photostat W.B.R.L.
(xii) W.B. Rent Roll, 1731, P.R.O.; photostat W.B.R.L.
(xiii) W.B.M.C.R., 2 Jas. II., P.R.O., transc. Lissimore, W.B.R.L.
(xiv) M. Willett, op. cit., p. 224.
(xv) S. Ryder, op. cit., W.S.L.
(xvi) M. Willett, op. cit., p. 220.

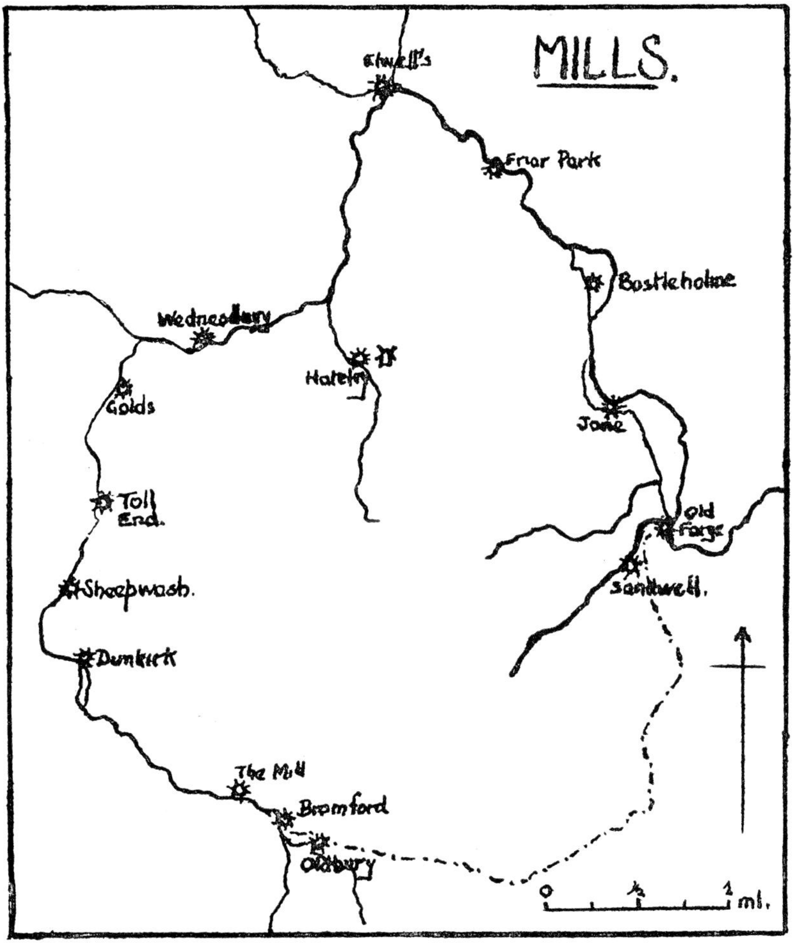
MILLS.
Elwell's
Friar Park
Bostleholme
Wednesbury
Hateley
Golds
Jone
Toll End.
Old Forge
Sandwell.
Sheepwash.
Dunkirk
The Mill
Bromford
Oldbury
0
½
1
ml.

West Bromwich Mills

IN an earlier chapter we have already noted the absence of mills in West Bromwich at Domesday. The building of a mill, important as it was to the life of the community, was never actually recorded so that we have no very exact knowledge of the beginnings of the numerous mills along the Tame valley. We can be fairly certain that they were small structures mainly of timber. Many of them must have been destroyed and rebuilt, possibly on slightly different sites, in the course of time. Though Wednesbury Mill was not in this Manor it was on the boundary stream and may be quoted. When, c. 1230 (i), William de Heronville demised this mill and mill suit to the abbot and monks of Bordesley, he gave permission for the said monks to transfer the mill from place to place within the same territory. Jone Mill appears to have suffered that fate for on the site it occupied early this century it could not have been served by a mill pool as it was in 1606 (ii).

Within the next century or so a number of mills were constructed. When William of Offney made his gift of Sandwell grounds to the Benedictine Order in 1180, he supported it by the donation of a mill at Greet (Sheepwash Mill?) (iii). At the Taxation of Pope Nicholas IV in 1293 (iv), the Abbot of Hales held in West Bromwich one mill worth 5s. per annum and the Prior of Sandwell held two mills to the annual value of 10s. The Abbey of Hales had been in possession of a mill at Friar Park for nearly a century. The other two were at Greet and Sandwell.

It would be surprising if a manorial mill had not been constructed by this time, but we have no direct evidence. There is no evidence of the legal disputes between tenants and the manor of mill suit such as frequently occurred elsewhere, as at Wednesbury in 1287 (v). Was this due to the non-existence of a manorial mill or to a more law-abiding tenantry and a more tolerant Lord of the Manor? The name of William le Molewart occurs in the Poll Tax Roll of 1379. This appears to be the first confirmation of the existence of the manorial mill.

Early in the 16th century the suppression of the monasteries was heralded. In 1526 an inquisition was held as to the possessions of the Priory and mention is made (vi) of a mill in West

Bromwich. Hackwood (vii), quoting Plot, describes the Priory Mill at Sandwell as ' a water-mill on the stream side, built with good substantial timber, 20 ft. by 15 ft. which was wont to go by the water of the pool which be now decayed and which with a little cost could be made to go again, for it hath sufficient water belonging to it if the heads of the pool were mended '. These properties were in the hands of Robert Whorwood when he died in 1590 and of his son, William Whorwood, who died in 1614.

By the end of the 16th century there was a string of mills along the Tame valley, in addition to the manorial mill which was almost certainly established at Hateley before this time. We have records of Joan Mill from 1526, Dunkirk Mill from 1580, of Bromford Mill and ' The Mill ' from 1583 and Bustleholme Mill from 1594. Forge Mill does not appear as such until the first decade of the next century but almost certainly existed slightly earlier than this. This makes a total of nine mills within the bounds of West Bromwich with others outside, but sharing the stream on its further bank. Undoubtedly they were mostly corn mills at some time with the exception of the Old Forge and probably Bromford Mill. There were no Fulling Mills in West Bromwich or Paper Mills so high up the river as this.

This is a surprising number when one considers the small area of arable land and the small amount of water flowing in the Tame. Undoubtedly, the competition caused friction. Frequent orders were made in the Manorial Courts as to the clearing of ditches and waterways (viii) and forbidding the diversion or damming of streams. As might be expected the manorial mills were specially protected. It was ordered in 1686 (ix) that ' no one may prevent the water within this manor flowing towards the lord's mills within the same manor when those mills shall need the same water ' under penalty of 5s.

In the second half of the 16th century the nailing industry was arising in West Bromwich. The Turton family who settled here about 1570 were among the earliest nailers. In 1583, William Turton the elder and Thomas Piddock, alias Cooper, of Oldbury, bought two existing mills from John Skevynton (x). They were together on the site of ' The Mill ' in Oldbury Road. The one was already a blade mill. In 1594 (xi) the Bustleholme Mill was purchased by Walter Stanley, Lord of the Manor of West Bromwich, from the Comberfords of Wednesbury. It is uncertain when it was converted to iron work. Woodman and Reignolds were lessees of two water-mills under one roof in 1625 (xii) but whether it was in their time or during the lease of Peytoe and Fowkes between 1629 and 1650 cannot be ascertained.

Foley, another iron master, joined Fowkes in 1649 (xiii). By

1630 it was known as the Rod Mill (xiv), a name which persisted until the end of the 18th century. In the 1626 Rent Roll we find, at that time 'in hand', 1 windmill and 2 watermills, i.e., the manorial mills at Hateley. The fact that there was a windmill suggests that the Hateley Mills had already gone over to iron work. That same year, Richard Foley took over the watermills at Old Forge, establishing a forge, hammer mill and furnace. The furnace was in such need of repair that he had it rent free for six years (xv). There was probably insufficient water power available on this site for Foley moved his activities to Bustleholme in 1649. The Hateley Mills reverted to a forge and finally became a foundry under such people as Thos. Webb, Ed. Elwell and Ed. Reddell until it ceased to exist at the end of the 19th century.

It is perhaps surprising that, despite the rising demands of industry, such a close agricultural connection should be retained by the mills. Even Bustleholme was never completely absorbed in the iron trade. The occupants or sub-tenants throughout the 17th and 18th centuries ran farming activities alongside. In 1709 (xvi) when John Lowe bought it from John Shelton, Lord of the Manor, it was described as one corn and one slitting mill together. In 1732 (xvii) when John Lowe conveyed it to his son, Jesson Lowe, it was a rod mill and oil mill together. It was part of the iron trade for the last time when leased by Thomas and William Morris of Bradeley in 1819, but reverted to flour milling under George Dorsett, who occupied it from 1826 to 1846 (xviii). It is doubtful if 'the Mill' was part of West Bromwich iron industry, from about 1610, while powered by water. Certainly the Turtons lived there from 1592 to 1735, but there is no evidence of it being used in that trade, though they were primarily nailers. It seems probable that they concentrated such work on the Bromford Mill from 1610 and purchased much of their rod from outside sources (xix). In 1708 (xx) there was a deed which referred to its as Josiah Turton's corn mill.

When Josiah Turton died it passed to his daughter, Margaret Abney. It was still known as Abney's (late Turton's) Mill in 1777. At the death of Mrs. Abney the estate was in chancery and the mill was advertised for sale (xxi) in 1782 as 'Lands and water-mill. To be let from Lady Day next, the mill estate lying in West Bromwich and Oldbury (about five miles and a half from Birmingham, the same from Walsall and three from Dudley) containing about 75 acres and a set of overshut water corn mills thereon, now in the tenure of Mr. Thomas Hawkins. N.B. The above set of mills stand very near the Birmingham Navigation and are now used for grinding corn and dressing leather, and are in an eligible situation for these purposes or for the Birmingham Manufactories'. It was purchased by Izon and Whitehouse, of Birmingham, manufacturers of hinges and small iron goods (xxii) who, probably about 1785 or 6, removed the water wheel and purchased a second-hand Boulton and Watt steam engine. The firm of Izons still occupies the site.

The Dunkirk Mill was another which played little part in the iron trade. Robert Ryder came to live there before 1580. In numerous deeds and in the parish registers the occupation of the family is always described as that of miller, not nailer. As they prospered they moved to Dunkirk Hall nearby and employed others to work the mill. In 1609 (xxiii) Simon Ryder employed as 'mylner' one Owen Davies at a wage of 45s. per annum and provided him with a shirt, soap, thread and lingels (leather straps). Davies was to work the mill and be responsible for its upkeep but was not to be called on for other work except at harvest time. In 1625 (*idem*) Simon Ryder paid John Amerie £6 15s. 4d. for a pair of millstones. The last of the Ryders, Nicholas Ryder, died in 1703 and the estate, heavily mortgaged, passed into the hands of three trustees and remained in chancery for nearly 100 years. Throughout this time the Gutteridge family worked the mill as a flour mill. The family appears in the West Bromwich Parish Registers from 1721. By 1836 it had at last given way to the iron trade for Reeves (xxiv), who states that the Gutteridge family had lived upon the estate for upwards of 100 years, refers (xxv) to Mr. Spencer's Mill, formerly Gutteridge's flour mill, as an example of the 'old school' of iron making, where the hammer is lifted by a water wheel.

It is obvious that the construction of the canal had produced little effect on this mill. A rental of the Dunkirk Estate in 1838, accompanied by a plan by Cooksey (xxvi) refers to Dunkirk Forge, stream, road and water power apparatus belonging thereto. A bundle of plans (xxvii) show the forge and mill dam at different dates from 1838 to 1868.

The Friar Park Mill had a long life. After the Dissolution of Hales Abbey in 1539 it passed to the possession of Sir John Dudley who converted it to a Bloom Smithy before 1553. It never reverted to corn milling and Reeves (xxviii) states that it was disused by the early 19th century, the building being in ruins in 1836.

The 'Old Forge Mill' had no agricultural phase prior to the end of the 16th century when a forge and furnace were built. The Rent Roll of 1626 names both furnace and a hammer mill. By 1836 Reeves (xxix) stated 'the forge has been in ruins several years and the mill used for slitting iron, about a quarter of a mile distant, is now a corn mill. They are part of the Sandwell estate'. The mill continued to grind corn into the first decade of this century. One field is still known as Hammer Mill Meadow. The

Sandwell Mill, on a small un-named tributary, could never have been very important after the dissolution of the Priory. An attempt was made to make use of the site for industry, for on Yates' map of 1775 it is named as a slitting mill. Later it was converted to a saw mill for the census returns of 1851 name the sawyers living there. It was marked as a 'saw mill' on the first edition of the 6 in. O.S. map, but the site was buried under the colliery spoil heap at the turn of the century.

Windmills never played a large part in West Bromwich. Mention has been made of the windmill at Hateley, which was probably built towards the end of the 16th century when the adjacent manorial water mills were leased for iron work. It continued to function as a corn mill at least until 1855 when Jn. Hewkins was the miller (xxx).

Two other windmills were built in West Bromwich, one in Lyndon Field close to the church, and the other near to Sandwell Road on the edge of the heath. Both of these were corn mills and both latecomers being built at the end of the 18th century. They continued to operate for over fifty years despite competition from steam mills which were later built.

(i) S.H.C., Vol. VI (i), p. 170.
(ii) W.B.M.C.R., 4 Jas. I., P.R.O., transc. Lissimore, W.B.R.L.
(iii) M. Willett, *History of West Bromwich* (1882), pp. 145-146.
(iv) Stebbing Shaw, *Natural History of Staffordshire* (1798) I, p. 50.
(v) F. W. Hackwood, *Wednesbury, Ancient and Modern* (1902), p. 33.
(vi) S. Shaw, op. cit., p. 128.
(vii) F. W. Hackwood, *History of West Bromwich* (1895), p. 26.
(viii) W.B.M.C.R., 16 Eliz., P.R.O., transc. Lissimore, W.B.R.L.
(ix) W.B.M.C.R., 2 Jas. II., P.R.O., transc. Lissimore, W.B.R.L.
(x) S.H.C., Vol. XV, p. 149.
(xi) M. Willett, op. cit., p. 184.
(xii) W.B. Rent Roll, 1626, P.R.O.; photostat W.B.R.L.
(xiii) W.B. Rent Roll, 1649, P.R.O.; photostat W.B.R.L.
(xiv) W.B. Parish Register, Parish Register Soc.
(xv) Footnote to Rent Roll for 1626, W.B.R.L.
(xvi) M. Willett, op. cit., p. 187.
(xvii) M. Willett, op. cit., p. 188.
(xviii) Sundry Deeds Coll., W.B.R.L.
(xix) *Foley Account Books,* Hereford County Record Office.
(xx) M. Willett, op. cit., p. 201.
(xxi) Aris's Birmingham Gazette, Oct. 7th, 1782.
(xxii) W. E. Jephcott, *House of Izons* (1948), p. 46.
(xxiii) S. Ryder, *Hypomnema,* W.S.L.
(xxiv) J. Reeves, *History of West Bromwich* (1836), p. 54.
(xxv) J. Reeves, op. cit., p. 114.
(xxvi) Tipton Reference Library.
(xxvii) Tipton Reference Library.
(xxviii) J. Reeves, op. cit., p. 113.
(xxix) J. Reeves, op. cit., p. 112.
(xxx) White's *Directory of Birmingham and District,* 1855.

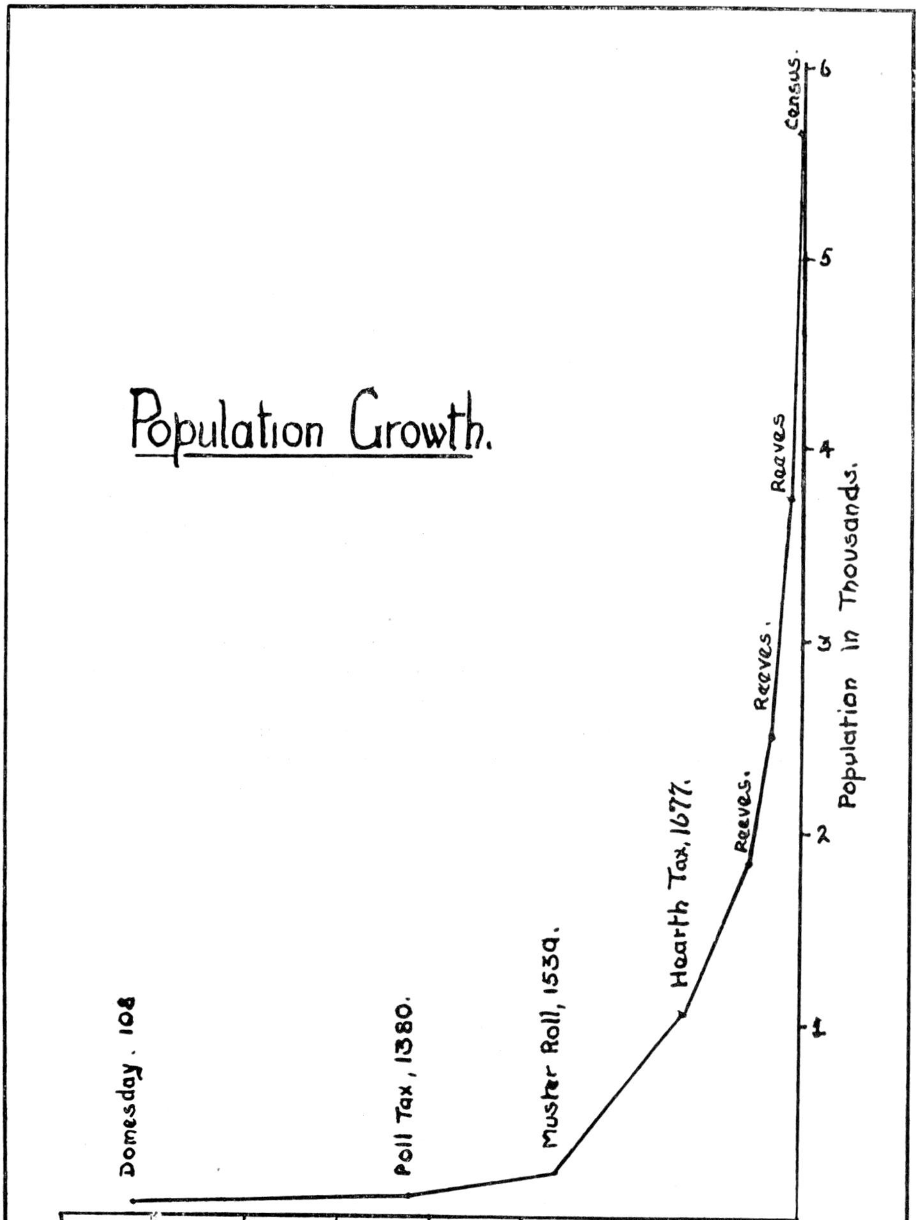
Population Growth.
Domesday. 108
Poll Tax, 1380.
Muster Roll, 1539.
Hearth Tax, 1677.
Reeves.
Reeves.
Reeves
Census.
Population in Thousands.
1000
1100
1200
1300
1400
1500
1600
1700
1800
1
2
3
4
5
6

The Arrival of Industry

THE second half of the 16th century saw a sudden rise in numbers and rate of growth of the population. Prior to this time the occupations of the people here had been agricultural or associated with the daily needs of a rural community. There had been the smith, the wheelwright and carpenter, the miller, baker and brewer. The bucklemakers, chapers, locksmiths and nailers already active in neighbouring towns had not yet arrived in numbers. At Walsall in 1538 ' there be many smithes and bytte makers in the town ' (i). William Turton who settled in West Bromwich about 1570 must have been one of the earliest of the nailers here. At this time, due in part to enclosures, there were many destitute wandering the roads of the country only to be driven on from place to place. Few would welcome them lest they became a charge on the parish. In 1573 an effort was made to restrict incomers and to reduce the risk. An order (ii) ' is made that if anyone shall have received into their houses any subtenants called Inmates without first having and obtaining a licence from the Lady (of the Manor), and six of the aforesaid jurors, they shall remove them before the Feast of St. Michael the Archangel next coming '. ' Forfeit for every offence, 20s.'

This did not stop the immigration and a number of fines were levied on those who gave shelter.

Thirty years later the picture had changed; the mills had turned to the iron trade, rods were being split for nailing, the furnace had been established and the iron masters were firmly embedded in the life of the community. Early attempts were made to achieve standards of workmanship and to protect those within the trades. In 1606 (iii) three orders were made by the local Court Leet with these objects. The orders were that ' no Inhabitant within this Manor using the trade, mystery or occupation of a smith, naylor or bucklemaker shall set or let any stall or stock to any person or persons of the same trade or occupation to work upon; that none within this lordship shall take into service any stranger or foreigner to serve as a journeyman in any trade or occupation to be used within this manor if the same stranger or foreigner have wife or children, unless they have habitation here within this manor, nor shall retain any for less time than one whole year; nor shall he retain any such foreigner or stranger unless the party so retained shall first find pledges for his good behaviour according to the law; also that no inhabitant of this manor shall retain any person to serve as an apprentice otherwise than by indenture according to the law and shall at the next Court Leet to be holden within this manor, bring or cause to be brought such Indenture to the Steward of the same Court Leet for the time being, nor shall he assign, sell over, release or part with any such apprentice so retained before the end of his term, without the consent and allowance of two Justices of the Peace next adjoining within this county '.

One wonders whether these regulations had any long-term effect for the invasion had begun. The incomers were to pour in, in ever increasing numbers, claiming ' squatter's rights ', and starting a trade which called for no land, little capital and minimal skill. In 1687 (iv) there were 71 people fined for erecting cottages, usually miserable hovels, on the margins of the heaths, and in 26 cases they had built a small workshop alongside. The figures for the next year were 66 cottages and 25 workshops.

This was the beginning of the industrial phase which was to so completely change the appearance and character of West Bromwich.

(i) Leland, *Itinerary* (1538).
(ii) W.B.M.C.R., 16 Eliz., P.R.O., transc. Lissimore, W.B.R.L.
(iii) W.B.M.C.R., 4 Jas. I., P.R.O., transc. Lissimore, W.B.R.L.
(iv) W.B.M.C.R., 3 Jas. II., P.R.O., transc. Lissimore, W.B.R.L.

POSTSCRIPT

by Dr. J. M. Fletcher, President, The Black Country Society.

Since its formation in 1967 the Black Country Society has published a number of books and pamphlets relating to local history. At present the following are available:—

Freeman's Black Country Folk (an important study of the Bilston area at the turn of the last century) 25p plus 4p p. and p.

The Stourbridge Glass Industry in the Nineteenth Century, by H. J. Haden, 30p plus 4p p. and p.

Tramways of the Black Country, 30p plus 3p p. and p.

History of Tipton Parish Church, 20p plus 3p p. and p.

Many articles relating to local history are also available in *The Blackcountryman,* the quarterly magazine of the society. For details of these and other publications, and also for general information about the society, membership, and subscriptions please contact the membership secretary:

Mr. Derek Simpkiss,
4, Stokes Avenue,
Portobello,
Wolverhampton WV13 3JT
Tel. Willenhall 67352.